Statistics: Short and Simple

By

James Jeray

Green Bay, WI

ISBN: 1-4107-8371-5 (e-book)
ISBN: 1-4107-8370-7 (Paperback)

Library of Congress Control Number: 2003095856

This book is printed on acid free paper.

Printed in the United States of America
Bloomington, IN

1stBooks – rev. 09/15/03

Table of Contents

Acknowledgements

I would like to thank Tracy Opicka for her advice, support and encouragement, also Marilyn and Marc Jeray for editing and other assistance.

"There are three kinds of lies: lies, damned lies, and statistics."
        —Mark Twain.

Introduction

Most people seem to have the same negative impression of statistics that Mark Twain apparently had. They don't understand it, don't trust it, and when pressed will readily admit that they are either intimidated by it or have no interest whatsoever. When I admit to having taught evening statistics classes to business students, looks range from awe to horror. People hope I will not say something that will magically give them a headache. This statistics-phobia seems to be quite common.

How many people have I seen walk into my statistics class with that "deer in the headlights," "what have I gotten myself into" look in their eyes? How many had left it until the last semester before graduation? How many would have skipped it entirely if it were not required? These are people who are not the least bit intimidated by batting averages, third down completion percentages, playing the lottery or going to the casino, basketball free throw shooting percentages, or even infant mortality rates. Yet the very thought of statistics as a subject of study sends shivers down their spines.

Others have said to me, "I would really like to know more about statistics, but I don't want to spend the time taking a course and working through all of the problems at the end of the chapters. People I work with who understand it seem to have an advantage. I just want to understand more about it." Most don't know that it's possible to learn just a little without getting bogged down in the complexity.

This short book is intended to address these problems, by describing the concepts of statistics without requiring calculations, by giving readers an idea of what it's all about in a friendly and accessible way. The organization is similar to most basic statistics text books, but the emphasis is changed from formulas and calculations to ideas about what is happening and why. It is more important to know *what* a standard deviation is and *why* you might want to calculate one than it is to know *how* to calculate one. A simple spreadsheet program on a PC can do the math.

We see almost nightly on the national news stories about what new drugs are being tested, what foods have been shown to be harmful, what vitamins may prevent cancers, what health practices will allow you to live longer, and what additive that was once assumed to be bad for you has now been shown to be not so bad. Unfortunately statistics like these are all around and people need to know enough to deal with them. Both in business and in our personal lives knowing about statistics is often the only way to tell what is important from what is just the latest fad. Without knowing something about the subject, people are exposing themselves to the chance of being confused, or worse, being taken advantage of, by those who can use or misuse statistics to their own advantage. To put yourself in a stronger position you don't need to know how to do any calculations. You just need to have an understanding of the subject, its strengths and weaknesses.

Being familiar with the concepts will make it easier for those planning or required to take a formal course later. It will also make it easier to survive the daily onslaught of news headlines designed to get your attention about the "latest research" through fear or exaggeration. It will help people at work dealing with programs like total quality management or six sigma, all of which are based on statistics. I used to tell my students that if they only knew the concepts and the vocabulary they could always remain calm and look into the details later. If someone at work mentioned standard deviation or chi-square and saw that lost look in their eyes, they were already at a disadvantage in the conversation.

Although most of the math involved in a basic statistics course is no more difficult than algebra, it can serve to distract and confuse people. Students memorize formulas and miss the point. They learn to recognize patterns of problems, so that they can "plug and chug" through the calculations. At the end of the course most people forget the formulas that they crammed into their heads for the exam and are left with nothing but the good feeling of having finally passed the course.

In summary, there is a large number of people who are exposed to statistics either formally through college requirements or informally through daily references in the news, weather, and sports. Most are not fully equipped to handle this information. The current alternatives of taking a class or reading a standard text book are not practical.

Here's what lies ahead.

The first chapter describes the need, how statistics are all around, unavoidable, and surprisingly so common that most people take the situation for granted.

Chapters 2-7 walk through the ability to use statistics descriptively, using simple examples and charts to show the basic concepts. It begins with some key definitions followed by examples of some "data visualization" techniques. Subsequent chapters explain the mean and standard deviation, distributions, again with simple examples and figures to illustrate various points. Chapter 5 on probability (a must for people interested in winning the lottery) is designed to provide comparisons between stated probabilities and everyday events.

Chapters 8-11 explain how statistics is used in surveys and studies to draw conclusions. This inferential statistics is at the core of reports on scientific findings. Questions like how to think about sample size and what is meant by plus or minus 3% in surveys are covered.

Finally Chapter 12 briefly describes some other common techniques and their uses.

Chapter 1 – The nature of the world

In my first class session on basic statistics, I would break the ice by asking the students to help me solve a personal problem. I told them that every semester when I came home from the first class session, my wife would ask, "How tall is your class?" How do I answer a question like that? I could lay all of the people end to end and measure the total height. One semester I might say that the class is 68 feet tall. The next semester I might say that the class was 111'2". The response might be, "Wow, are you teaching the basketball team?"

It was a trivial example because the heights of students is not a very important piece of information. But it is instructive and most of the students were at the point of recommending the <u>average</u> height before I finished the explanation. After all, everyone knew about averages. But an average is just a way to express with a single number an idea that began as a lot of individual numbers, each student's height. In the example above there may have been 12 people in the first class for an average height of 5'8" and 20 in the second for an average height of 5'7" – so it wasn't the basketball team after all.

This ability to take a large number of measurements and express it in an abbreviated way, as one or two numbers, is the key behind descriptive statistics.

Think about it. Almost everything in the world varies. Heights of students is just one example. Weight, age, birthday, blood type, life expectancy, hair color, intelligence, salary, house size, pulse rate, and everything else you can measure about the students will vary. In fact, you would be very surprised if it didn't vary. Suppose you looked at a painting of a forest and all of the trees were exactly the same size. You may suspect that it was done by a young child or done with a stencil, but it would not look realistic. Suppose that every time you stopped at the service station to fill your car with gasoline the number of gallons was exactly the same – even twice in a row might strike you as odd. Suppose that each visit to the grocery store cost exactly the same when rung up at the cash register – again even twice in a row would seem unusual. Suppose that the lottery numbers came out the same two days in a row. That would make headlines! People who work with machinery that cuts or shapes know that even these machines don't cut every piece to <u>exactly</u> same length. If you can measure with enough precision, you can always detect some variation.

The point is that almost everything in the world varies, and we are so used to it that we don' t usually think about it until we run across a case where it doesn't. If the electric bill or phone bill is exactly the same as last month, then it strikes us as unusual and we may even get suspicious. "Something is wrong."

Since so many things are variable we need a way to cope with it. First we need a way to describe things. We need some shorthand way to talk about this variability without having to list all of the numbers. **Descriptive statistics** uses one or two numbers to summarize the set of all measurements in a way that gives you some, but not perfect, information about the whole set of numbers. The average height says something about all of the measurements of the individual heights of students.

The other problem with variability arises when trying to decide if a measurement fits or does not fit with another group of measurements. The recovery rate from a certain illness will vary from patient to patient. Some will just improve faster than others. If you give some patients a special drug, do they really improve faster because of the drug, or is it just the result of natural variation? Do the treated patients differ by a large enough amount from similar untreated patients to prove that the treatment works? Is it a miracle cure or just a fluke caused by the particular people chosen for the clinical trial? How do you tell? This is the type of issue addressed by **inferential statistics**.

The need for statistics arises from the natural variation of measurements in the world. The first step is to collect the data. When people must know something about a large number of measurements, knowing that they vary, they use statistics as a tool to describe the group, or they use statistics to test whether there is truly a difference between groups.

Before getting too far along, it will be helpful to define a few commonly used terms. Then in Chapter 3, I will give examples of how to make information easier to understand just by the way it is displayed, making data into pictures. Chapter 4 will begin the discussion of descriptive statistics.

Chapter 2 – Some Definitions

In statistics people use words in a special way. Sometimes this is where the confusion arises. Below is a list of common terms along with a simple definition of each. Examples are added where it may help to clarify.

<u>Population</u>: the whole area of interest. What are you making the measurements on? Is it the students in a single class, the students in the whole school, or all students in the country? Is it the output of a certain machine? Is it the products manufactured by a certain company? Whatever the area of interest, all possible items that could be measured make up the population.

<u>Sample</u>: Often it is not practical to measure the entire population (see census). Then only some of the members of the population are measured. When survey companies or polling groups want to know how all voters in the United States feel about an issue, they generally ask only a little over 1,000 people. It is too expensive to ask everyone, and 1,000 usually gets a pretty close answer. When a company wants to advertise how long their light bulbs will last, they burn a few until they burn out. If they tested them all, they wouldn't have any light bulbs left to sell!

<u>Census</u>: Every ten years the government tries to do a census. This is a measurement of certain traits of the entire population. It is usually difficult and expensive, and not practical for everyday questions, so sampling is used far more often. If a company wants to know how morale is, they may send everyone a questionnaire in an attempt to conduct a census rather than sampling. (The population is all employees.)

<u>Variable</u>: This is whatever you are taking a measurement of: height, weight, age, income, blood pressure, color, or acidity. The variable is assigned to a <u>subject</u>. Here is a list of items with a subject followed by a variable: student/height, tree/age, apartment building/number of units, house/square feet, parking lot/number of spaces. The variable is always something that can be counted or measured.

<u>Observation</u>: Each individual measurement is considered an observation. The height (variable) of the first student (subject) is 69 inches. This a single observation.

<u>Continuous vs. Discrete Data</u>: In simple terms, if a measurement can increase smoothly, like the height of anything that grows, it is continuous. If a measurement jumps from one whole number to the next, like the

number of people in a room, it is discrete. Some radios have a volume knob that turns in a smooth motion from low to high – continuous. Other radios have a volume knob that goes "click, click" from one setting to next – discrete.

The reason this distinction is important is that different techniques are used with each of these two types of variables. The famous "bell-shaped curve" usually applies to continuous variables.

Mean and Standard Deviation: These are commonly used terms to describe the center of the measurements and how spread out they are. The mean is the same as what we learned in school as the average. The standard deviation is a special measure of how different the measurements are from each other. (More on this in Chapter 4.)

Parameters vs. Statistics: Characteristics like mean and standard deviation are called parameters when they are measured on the whole population. They are called statistics when measured on a sample of the population. Greek or Capital English letters are used for parameters and small letters are used for statistics. For example "N" is the number of observations in a population and "n" is the number of observations in a sample. "Sigma" (σ) is the standard deviation of a population and "s" is the standard deviation of a sample.

Outliers: an outlier is an observation of a variable whose value is very much different from the rest. For example, a factory is built and a small town develops around it where all of the workers live and shop. Most of the people are working people in the factory, shops or small offices. But the owner of the factory also lives in the town. His house is 20 times the size of the average worker and his salary is 20 times the size of the average resident in the town. The house size and salary of the owner are outliers compared to the rest of the population of the small town.

Median: Like the mean, this is a way of expressing the central value of a set of measurements. If a statistician wanted to calculate a representative number for the size of a house in that same town, with most around 1000 square feet and one mansion at 20,000, the average would be "skewed" by adding the owner's house in because it is as big as 20 normal houses, but only counts as one. The mean or average would be a medium sized house. But the town has only small houses and one very large one, so the mean which is supposed to give reliable summary information about what you might expect to observe does not represent the size of houses in the town. In this case the statistician uses a measure called a median. It is calculated by taking the size of all of the houses in the town and arranging them from smallest to largest, then picking the middle house. The measure of this observation does represent a real house, and one that you might expect to find in the town. That is why, especially in economic reports on the news, you will hear the "median income" or the median of other variables. It is used to keep the outliers from unfairly influencing or skewing the answer.

James Jeray

Chapter 3 – Presentation of Data

Sometimes it is easy enough to show what data looks like without using any statistics to summarize it. You can display data in the form of a picture. Take, for example, the list of heights (in inches) of 12 students in a class.

Heights	68	70	66	67	63	68	70	69	67	64	68	69

Table 1

If this spreadsheet display of only 12 heights is not very easy to picture, imagine a business with hundreds or thousands of observations of some variable. An array of thousands of numbers is impossible to comprehend all at once.

A simple way to picture such data is by putting the heights on a graph as shown below in Figure 1.

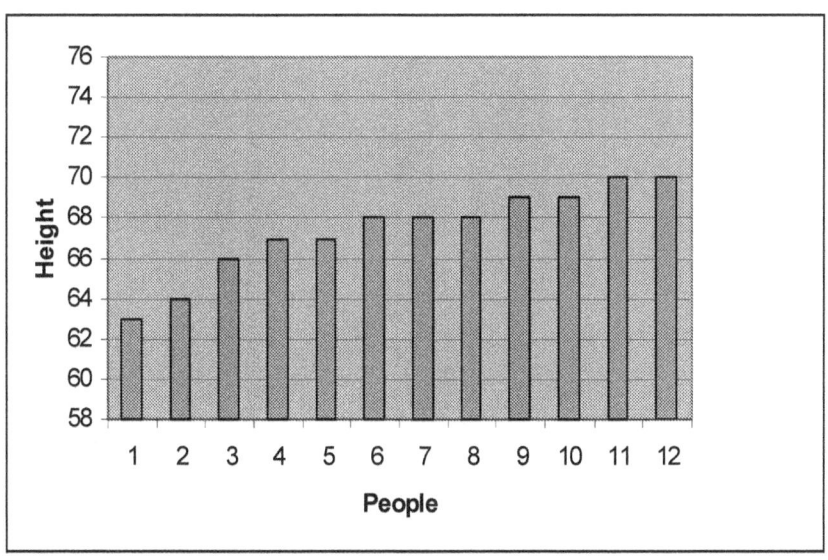

Figure 1

Now it is easy to pick out the tallest and shortest and to guess at the average. The same is often true with very large data sets.

Probably one of the most common examples of picturing data is called the histogram. For a each observation, the value of the variable is recorded as a solid block, and where there are repeat values another block is stacked on top of it. Shown in Figure 2 is a simple histogram of the heights of the same 12 students.

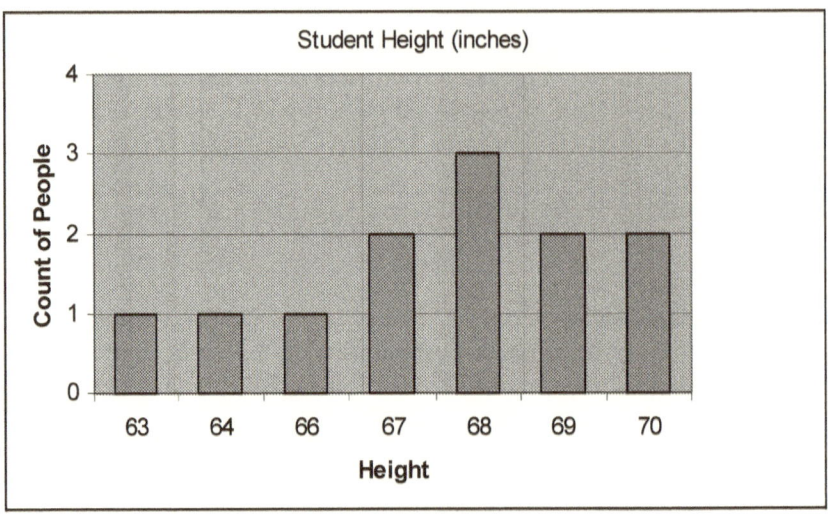

Figure 2

This gives the same information in a slightly different way.

Another way to show the same data is in a pie chart, as in Figure 3.

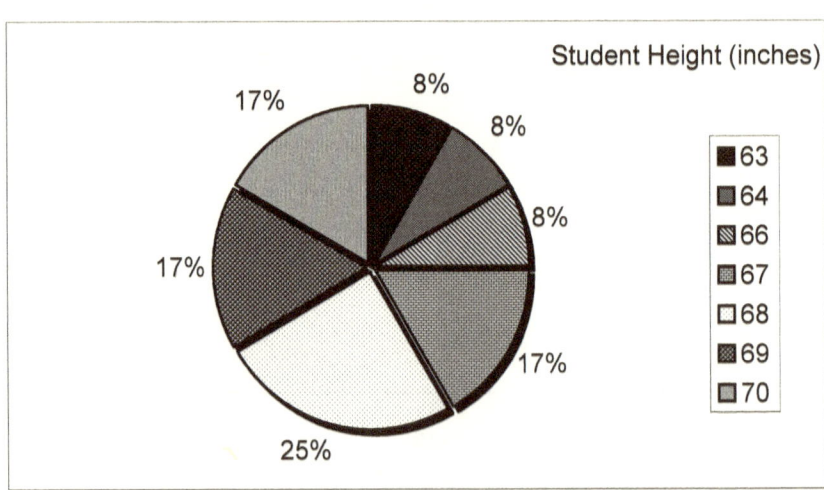

Figure 3

There are many choices of data presentation tools. Each gives an idea of the range of values, the approximate central value, and the amount of variation in the values. Some tools are more effective than others. It depends on the data and the purpose of the graphs. Each has its own strengths and weaknesses. As seen earlier, Figure 1 shows a good picture of the height of students because the graph looks like a bunch of students standing next to a ruler. The pie chart shows that there are more at 68 inches that any other height, but does not give as good of an overall picture. The histogram, as in Figure 2, is more commonly used in statistics.

Unfortunately there are many ways to give a wrong impression by using the wrong presentation tool or by using the right one improperly. For example, a first glance in Figure 1, student number 11 looks like he is more than two times the height of student 1. That is because the scale does not go down to zero. This presentation can make small difference look very large and give a false impression of the amount of variation.

The ways to deceive or give a false impression using these presentation tools are too numerous to cover. A good practice is to just look carefully, read the scales and be aware that the picture is merely a tool for summarizing the data and may not be totally accurate.

For more examples of these presentation tools, most of them very well done, pick up almost any issue of the USA Today.

James Jeray

Chapter 4 – Describing Data

The "pictures" of the data in the previous section are helpful as summaries because they answer two important questions. Where is the data located, and how spread out is it? Using the same heights of students example, I want to know if they are mainly short or tall; that is the *where*. I also want to know if there are a lot of people of different heights or are they all about the same; that's the *how spread out.*

The where (sometimes called central tendency) is usually expressed by the mean, or average. For example, the class is on average 68" tall. The average is calculated by adding up the measurements of each observation and dividing by the number of observations – just like a bowling average is calculated by adding the scores of each game and then dividing by the number of games. In baseball a batting average is the number of hits divided by the number of at-bats.

My Bowling Average	
Game 1	154
Game 2	148
Game 3	139
Total Pins	441 /3 = 147

Table 2

Shown below are examples of two classes of students that are on average 68" tall but with different standard deviations. The values are shown first in the table below. (They are arranged in order to make it easy to see the differences.) The one on top of Table 3 is not very spread out at all, the shortest person is 63" tall and the tallest is 72". The one in the bottom row has a much larger variation in heights, ranging between 60" and 75".

Less Variable	63	64	66	67	67	68	68	68	69	69	70	70	71	72
More Variable	60	60	61	64	65	68	69	69	70	71	72	73	75	75

Table 3

Now look at graphs of the same information in Figures 4 and 5.

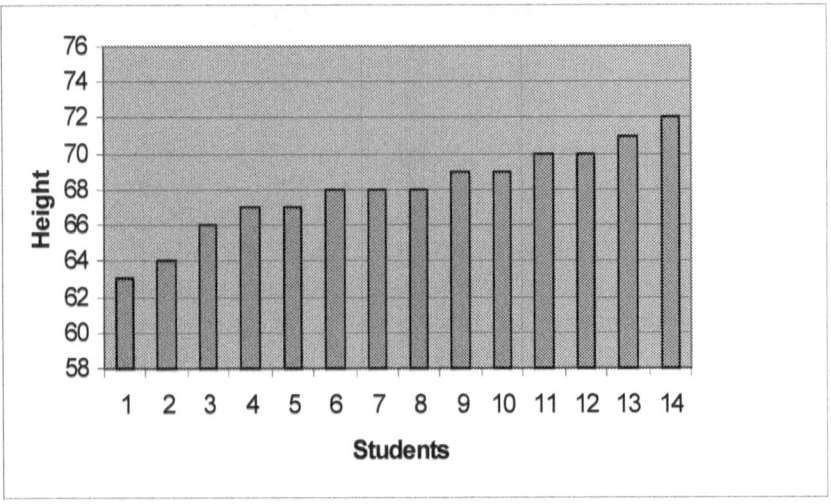

Figure 4

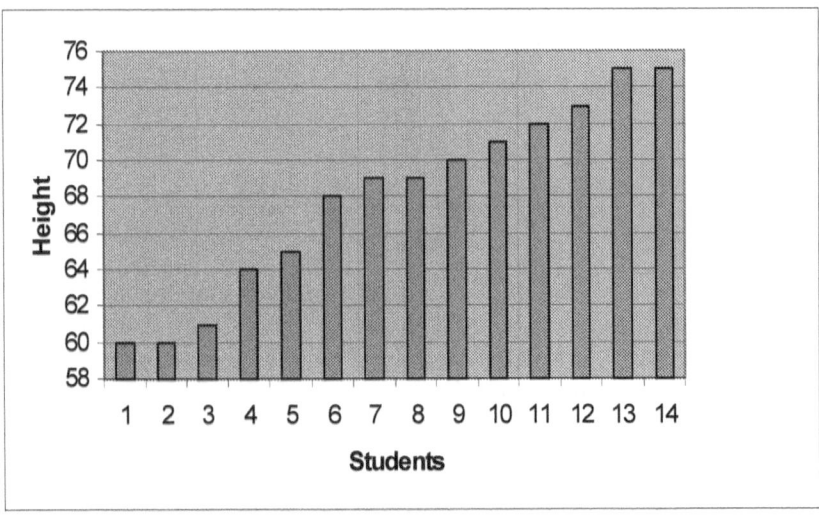

Figure 5

As mentioned before the mean of each data set is 68". That number describes the location of the data, where it is. It's all around the value of 68". The standard deviation for each though, is different to express how spread out the values are. The standard deviation of the first set is 2.51" while the standard deviation of the second is 5.23". This is a way to express with a single number how spread out the heights of students in each group are relative to each other. The larger standard deviation of the second group tells us that the heights are more spread out than the first.

The standard deviation is calculated by taking the distance of each observation from the mean, and squaring it, and adding them up, and then taking the square root. Why all of the squaring and square rooting? Some books explain that the distance can be positive or negative and the squaring makes them all positive. But there are other ways to make them positive, like taking the absolute value.

In fact there are a large number of ways to measure how spread out a set of data is. The standard deviation is only one of them. Many people are confused by trying to figure out what this means in a pure mathematical sense. Does this mean that the second is more than twice as spread out as the first (5.23 compared to 2.51)? What does twice as spread out mean? How does all of this squaring and square rooting work into the situation? A simple summary of the major points includes these three things.

- For the same kind of measurement, a larger standard deviation means the data set is more spread out.
- Any standard deviation is related to its own particular set of data, and you can't say in general that 2.5 is small and 5.2 is large. It just shows that, in this particular case, one is more spread out than the other.
- Despite what may seem like a contrived calculation, the standard deviation has some very handy uses that will be discovered later.

The examples given have been kept simple to allow better understanding of the concepts. In most cases where mean and standard deviation are used, much more data is involved. The value of these statistics lies in their ability to tell "where" and "how spread out" the data set is, especially when you don't have a simple table or graph.

Using the mean and standard deviation is one very common way to describe a large amount of information using only two numbers instead of trying to make sense out of hundreds or thousands of records. It is a kind of short-hand that says, "What you are measuring is about this size and the various (and variable) measurements cluster around it either tightly or relatively widely. In the height example it tells that people are all about this tall with either some much taller and shorter or with most about the same size.

The reasons for knowing this information applies to business and personal life. If someone is running a machine that cuts tubing for sale to a customer who makes pumping equipment and the tubing is supposed to be 5" long to fit into the equipment, there will usually be a limit on the variability allowed in the specifications. An average of 5" is the target for the cutting machine, but if the length varies from 4.5 to 5.5 inches the machine needs to be recalibrated. Likewise if the tubing is averaging 5.2" within a very narrow range, the company is giving away 4% more than they agreed to for the same price and is therefore losing money. They don't have to look at all of the measurements to discover the problem, just the two statistics, mean and standard deviation.

Suppose a football team converts on 35% of third down plays, but the standard deviation is quite high. The coach may discover that instead of being 35% across the board, it is 45% at home and 25% at away games. This might lead to some different coaching decisions for the upcoming game.

Knowing only the mean and standard deviation sometimes can help uncover the root cause of a problem.

Chapter 5 – Probability

When people talk about this subject it is often referred to as "probability and statistics." Probability is related to statistical in the way the results are expressed. Did you ever hear a report on the news about the latest political polls? If they are thorough they say that 53% of the voters favor an issue with a margin of error of plus or minus 3% at a confidence level of 95%. In the example of the football team in the last chapter, they gave a mean success rate on third downs of 35%, or about 1 in 3. For this reason it is important to have a basic understanding of probability to help understand some aspects of statistics.

Probability is familiar to most people in the form of the "odds" in gambling. What are the odds of throwing a 7 or an 11 in one role of the dice at the craps table in Las Vegas? In fact, much of probability theory was invented in response to an increased popularity of gambling in Europe a few centuries ago. As new games were invented to keep the king amused, people who liked to work with numbers so much that they didn't mind that calculators hadn't been invented yet, developed probability theory. Since the study of probability can take several college level classes, this chapter will touch only on the mere basics.

Probabilities are quoted, usually in terms of percentages, in many aspects of daily life. It may be the weather report telling of a 30% chance of rain, the small print on the back of a lottery ticket telling of a 1:55 million chance of winning, a sales manager reporting a 75% chance of winning a key account, a life insurance company calculating expected profits based on the probability of an accident among its policy holders, or an engineer discussing the likelihood of failure of a key component. These probabilities come from three main sources: calculation, sampling/testing, or an estimate based on experience. The weather forecaster is using an estimate to give the chance of precipitation. Examples of probabilities from calculation and from testing are discussed below.

The central idea of probability is to be able to state as a fraction or percentage the number of expected successes out of the number of total trials. Here is a quick example. Take out a coin and toss it. There are only two possible outcomes, heads or tails. Each toss is considered a trial. Pick heads as a "success." (Note that success does not necessarily have to be a good or favorable outcome; it is just the chosen result.) So for one trial, toss of the coin, there are two possible outcomes, and for a "fair coin" that is not weighted in any way, the chance (probability) of getting a head is 1 out of 2. There is one head out of two outcomes (head or tail). Common ways of expressing the probability for the outcome of a coin toss are 50-50, 1 out of 2, 50%, or ½.

By definition probabilities always fall between 0 and 1. If an event has a probability of 0, it will never happen. If it has a probability of 1, it will always happen. If the probability is between 0 and 1, it will sometimes happen, and the closer it is to 1, the more likely it is.

In future discussions, I will mostly use fractions to show probabilities. I find it easy to think of these fractions as the number of successes divided by the total number of possible outcomes. So the probability of getting a head is ½, because there are two sides to the coin and one of them is "heads." The probability of rolling a 1 on a standard six-sided die is 1/6. The probability of randomly drawing a heart from a deck of standard playing cards is 13/52 (or ¼), because there are 13 hearts and 52 cards in all. (In all examples assume that the coins and dice are fairly balanced, not bent or biased in any way and that these activities are done without cheating.)

The explanations above are examples where the probabilities are calculated. We will stick with calculated probabilities for a while before moving on to tested probabilities.

As long as examples are kept simple, probabilities can be calculated in a very straight-forward way. Even when you go beyond one flip of the coin or pick of a card, the probabilities are easy to discover. Two rules apply involving "and" and "or." For simple situations, these work fine.

The rules are:
- If it is a choice between one outcome "or" another, that is either is acceptable, then you add the probabilities for each event.
- If it is a combination of one outcome "and" another, that is success requires both, then you multiply the probabilities for each event.

This allows you to compute probabilities for multiple events like rolling the die and getting an even number. The probability of getting a 2 is 1/6, a 4 is 1/6, and a 6 is 1/6. Since any of these satisfies a success, you may get 2 <u>or</u> 4 <u>or</u> 6, then the probability of an even number is 1/6 + 1/6 +1/6 = 3/6 = ½. To check this we can go back to the original definition which tells us that the probability is successes / possible outcomes. There are 3 successes (2, 4,or 6) and 6 possible outcomes (1,2,3,4,5, and 6).

The "and" guideline helps in computing probabilities of compound events, like tossing two coins or tossing the same coin twice in a row. If you toss two coins expecting two heads, it is an "and" situation (calling for multiplication). Both results must be heads to satisfy the condition. So the probability is ½ x ½ = ¼. To take it further, the probability of tossing five in a row (or five coins at the same time) and getting all 5 heads is ½ multiplied 5 times, or 1/32. This is about 3%. So when someone says that there is a 3% probability of something happening, the likelihood is approximately equal to tossing 5 heads in a row.

It is important to be able to convert a given probability into terms that you are somewhat familiar with. Tossing 5 heads is about 3%, picking the ace of spades from a shuffled deck in one try is approximately 2%. Rolling a one on the die is between 15% and 20 %. Drawing a given suit from a deck of cards is 25%, and a coin toss is 50%.

From these you can look at the opposites. If the probability of doing something is a given value, then the probability of not doing it is 1 minus that same value. Why is this? Because the probability of a sure things is 1, or 100%, and to either get five heads or to not get 5 heads is a sure thing. If you toss 5 coins it is guaranteed that you will either get 5 heads or something else (not 5 heads) every time! So if the probability of getting 5 heads is about 3%, then the probability of tossing 5 coins and not getting all heads is about 97% (the remaining 31of 32 times)

Likewise the probability of rolling the die and not getting a one is 5/6 or between 80% and 85%. The probability of drawing a suit other than spades from a deck of cards is ¾ or 75%.

Just for the sake of perspective, the probability of going to the grocery store and having your total come out to an even dollar amount is 1/100, or 1%. There are ninety-nine other ways to come out to odd cents. So the probability of this happening on 3 consecutive trips to the store are 1/100 x 1/100 x 1/100 = 1/1,000,000! That's what one in a million is in common terms that most people can relate to. Can it happen? Of course. People win the Powerball lottery jackpot and the probability of picking these right numbers is over 100 times worse, so it is possible, just not very likely.

In some cases, especially where you are not dealing with coins, dice or cards, the probabilities can not be easily calculated. What is the probability of finding a defective part in a bin of identical parts? You don't know the successes, in this case defective parts in the bin. How does wearing a seatbelt improve your chances of surviving a traffic accident? How much more likely are you to be involved in a traffic accident, if you are drunk or using the cell phone or changing the station on the radio? The answers to these and many other similar questions can only be answered by gathering data over a period of time. The probabilities are then stated in the same way, successes over the total number of events. As the parts in the bin are used the number of defectives is counted and the total number of defective parts becomes known. If it is found that 30 parts out of 1000 in the bin are defective, it is assumed that 3% (30/1000) of parts from the same supplier will be defective in the next bin. Negotiations over the price of future shipments will use this information.

If this subject seems interesting to you, there are several short, fun books on probability that take the concepts to another level with many more examples. Check the library or bookstore.

In conclusion there are some important points to understand about probability.

- It is stated as a number between 0 and 1 with 0 = never happens and 1 = always happens.
- Probabilities can be expressed as fractions of successes / total trials, which may be stated as fractions but are often converted to percentage.
- In simple problems involving multiple activities, "and" means to multiply probabilities and "or" means to add probabilities.
- The probability of each separate trial is always the same and not dependent on previous trials. A coin doesn't remember that it was a head last time and now it is time to be a tail. Sometimes there can be long strings of heads in a row – but it does not happen often because it is unlikely. Don't be fooled into thinking that you can predict the next outcome based on the fact that it hasn't happened in a while. If a one has not been rolled on the die in a while there is exactly same chance of a one this time that there was last time, and that there will be next time.
- For learning statistics it is important to have some idea of what is meant by a 5% chance or a 3% chance, etc. a good reference for this is coins or cards or whatever makes sense to you.

Finally a word about beating the bank at Las Vegas (or any other gambling venue). You rarely meet people who will admit to losing money there, but the money to build those big hotels had to come from somewhere! The fact is that the payout on every game has a probability of winning of slightly less than 50%, a little less than breakeven. So over the long run, on average, they make a few cents on every dollar bet. Of course there is variation, so some people do come out ahead. But on average, every dollar that is bet contributes a few cents to the Las Vegas economy and not to your fortune.

Take a roulette game as an example. There are 36 numbers plus a 0 and a 00. That leaves one chance of winning out of 38, but a pay-out of only 35 to 1. This means that over the long run every 38 dollars bet costs the casino only 36 dollars (the $35 pay out plus the return of the winning dollar) which is about a 5% return to the casino. So every dollar bet has an expected value to the casino of over $1.05 and to the bettor of less than 95 cents. (Expected value is another term commonly used in statistics.) And remember, the roulette wheel has no memory, so if an 8 has not come up in a long time, that does not mean that an 8 is "due," or any more probable. The odds are the same for each of these independent trials.

State lotteries are much worse for the bettor. A pick-3 pays $500 against the chance of picking one correct number out of the 1000 numbers from 000 to 999. Picking that number correctly has about the same probability as tossing 10 heads in a row. It says right on the ticket that only about half of the money goes to prizes and the rest goes to tax relief and expenses.

So only play for fun. The odds are against you.

James Jeray

Chapter 6 – Common Distributions

Here, in Figure 6, is a copy of the histogram from the beginning of Chapter 3. It represents the heights of 12 students. It shows the range from 63 inches to 70 inches. Some bars are taller to indicate that there are more than one student with the same height. You can see at a glance that there are more students who are 68 inches tall than any other height. This idea of a range of values where some values have multiple observations within the data, is referred to as the distribution of the data. A distribution shows the location of the data, the range and where the most observations are concentrated.

A distribution is said to have a shape. The distribution in Figure 6 has a peak at 68 inches and seems to have a longer "tail" to the left, that is, toward the shorter people.

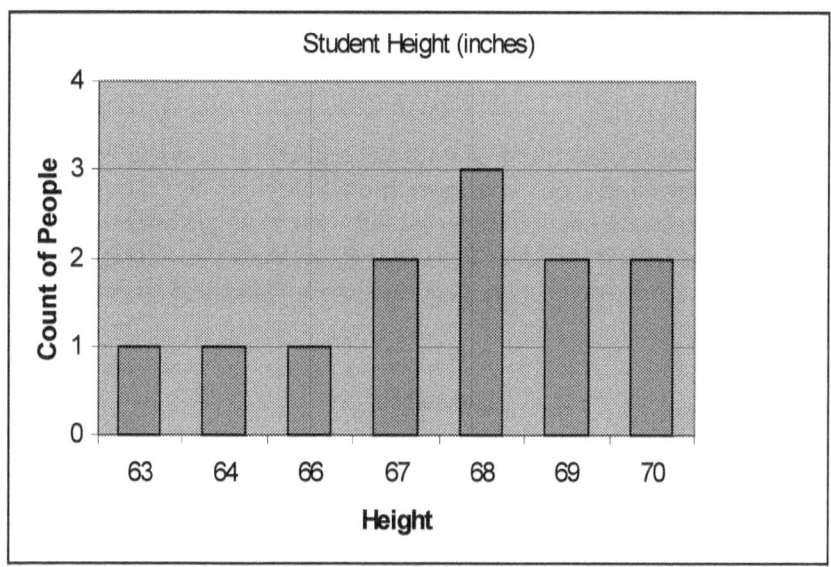

Figure 6

When there are many more observations than 12, the distributions get more densely packed, and the shapes seem to get smoother so that a graph of the distribution looks like a solid shape and is often shown as just a solid line. The point of this line is to show the height of the distribution at any particular point along the range. Remember, the height tells how many observations occur at this particular point. So the area under the curve holds much of the information.

The shape of the distribution reflects several terms that were covered earlier. The mean is the center of the distribution. The mean tells the location of the data and, therefore, the location of the distribution. The standard deviation tells something about the range, how spread out the distribution is. If you could look at a picture or graph of the distribution, the outliers would appear as lone points far from the rest of the distribution to one side or the other.

There are several common shapes of distributions including triangular, uniform, double peaked and sloped. Figure 7 shows a few examples.

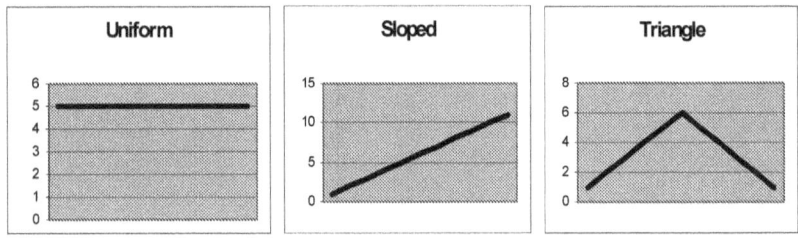

Figure 7

Many distributions appear to be smooth and peaked as shown in Figure 8. If the data from the observations are more spread out to the left (lower side of the distributions), it is said to be skewed left. The example in Figure 6 is slightly skewed left. If the opposite is the case as in Figure 8, it is said to be skewed right. If it is spread evenly on both sides of the peak, it is said to be balanced.

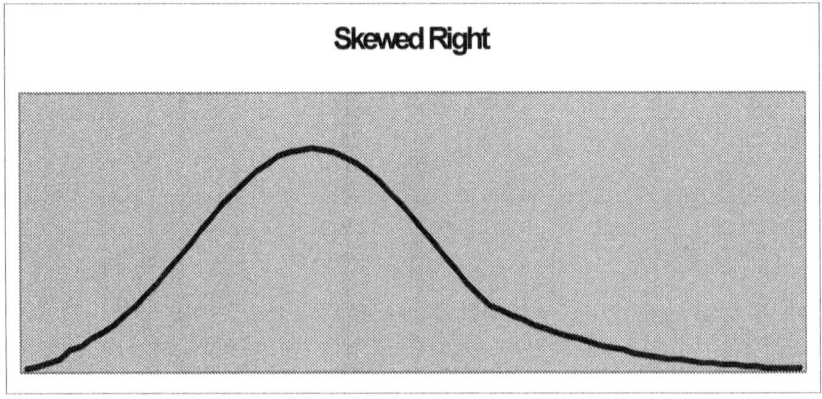

Figure 8

In your reading you may run across references to other distributions such as the F distribution, the exponential distribution or the Poisson distribution. These are all mathematically derived, have special characteristics, and describe or approximate certain natural occurrences. (They will not be covered here.)

When discussing distributions in general, two rules are introduced, the Empirical Rule and Chebychev's rule. They are rules that tell you something about a distribution even if all the information you have is the mean and standard deviation. They are as follows:

Empirical Rule: If the probability distribution is bell-shaped, then

> Approximately 68% of the observations will fall within one standard deviation on either side of the mean, and

> Approximately 95% of the observations will fall within 2 standard deviations of the mean.

Chebyshev's Rule: no matter what the shape of the distribution,

> At least ¾ of the observations will fall within 2 standard deviations of the mean, and

> At least 8/9 of the observations will fall within 3 standard deviations of the mean.

This shows the special nature of the standard deviation, and why it is often chosen as a measure of spread or variability.

If you are inclined to say, "So what," remember that the intention of descriptive statistics is to use only a few key numbers to describe a large set of data. In this case given only the mean and the standard deviation, Chebychev's rule allows anyone to make assumptions about data, the observations or measurements, in a distribution, regardless of its shape. Knowing the approximate shape allows those assumptions to be even more precise by using the empirical rule. Furthermore, knowing the exact shape allows far more precision about the data.

Normal Distribution

One common distribution is called the normal distribution, also known as the bell curve or bell-shaped curve. As it turns out, it can be used to represent many instances of real life measurements. The normal distribution has properties that make it very convenient to work with.

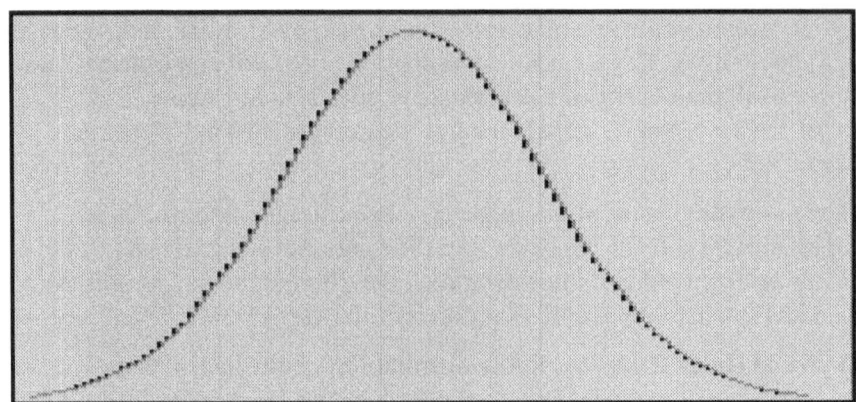

Figure 9

The power of the normal distribution is in its shape. It is balanced, not skewed in either direction. The mathematical formula for that line is the same for every normal distribution that exists. The only things that change are the mean (where it is located) and the standard deviation (how peaked or spread out it is). So although the different normal curves may look different, they can really be treated as all the same by artificially moving their center to 0 and treating their standard deviation as if it were equal to 1. All of the measurements can then be found in a table in the back of any statistics book. The general answer is then converted back to apply it to the situation in question. Here's how it works.

Since it is a continuous distribution, a smooth line, it is treated like it has no individual distinct points. Remember in the earlier examples it was possible to ask, "how many students were 66 inches tall?" In the case of truly continuous data the 66 inches blends immediately with 65.99999999 inches, so theoretically the divisions become infinitely small. That's why a normal distribution table refers to the area under the curve to the left or right of a certain point and not to the height *at* that point. This is a fine, but sometimes confusing distinction.

To convert a normal distribution to the standard which allows for a look up in the table, subtract the mean, which brings the new mean to 0, and divide any distances away from the mean by the standard deviation, which expresses the distances in terms of a unit standard deviation. This changes any normal distribution to a standardized normal distribution like the one in the table. The number given in the table will usually be the percent of observations to the right of the point in question or between the mean and the point in question. A picture usually accompanies the table to illustrate this point.

Here are a few examples. Suppose there is a large set of data with a <u>mean of 100 feet</u> and a <u>standard deviation of 10 feet</u>. It could be heights of tall trees in a particular location. To find how much of the data is less than 100 feet you look up 0 in the table (100-100=0). It will tell you 50%. This is pretty trivial since 100 is the mean, so by definition, 50% of the data is greater and 50% is smaller. This is shown in Figure 10 below.

Another person might wonder how many of the trees were taller than 110 feet. This is also shown in the figure below. 110 feet is above the mean of 100 so fewer than 50% of the trees are taller. The percentage is represented by the area under the curve. See how most of the area is massed around the center. Most of the trees are close to average height. As you move to the values away from the average, it becomes less likely to find trees that size. The table has all of the answers provided you can express the distance away from the mean in terms for how many standard deviations. This is called the z-score. (At one standard deviation z=1).

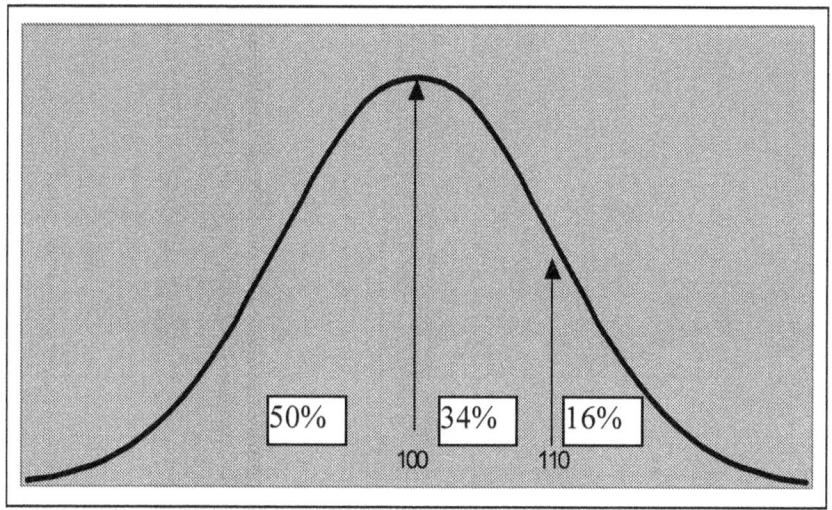

Figure 10

The point representing 110 feet is exactly one standard deviation, 10 feet, away. The table will tell you that about 34% of the distribution is between the mean and the point at one standard deviation away. Then it is just a matter of addition and subtraction. There are 50% shorter than 100 feet and another 34% between 100 and 110 (for a total of 84%). So that leaves only 16% taller than 110 feet.

It is very convenient that once you have a normal distribution with a known mean and standard deviation, all questions can be answered using the table and arithmetic.

Figure 11 shows one more example, this time of a tree that is smaller than average. In fact it is quite a bit smaller at only 85 feet. This is 15 feet or 1.5 standard deviations below the mean of 100 feet. (Remember one standard deviation, in this example equals 10 feet.) To answer the question of how many trees are no larger than this tree, the table is used again. This time the table says that the area between that point and the mean equals about 43%. Since the whole left half of the distribution contains 50% and the part between there and the mean is 43%, then the trees that are smaller, that is to the left, are the remaining 7%.

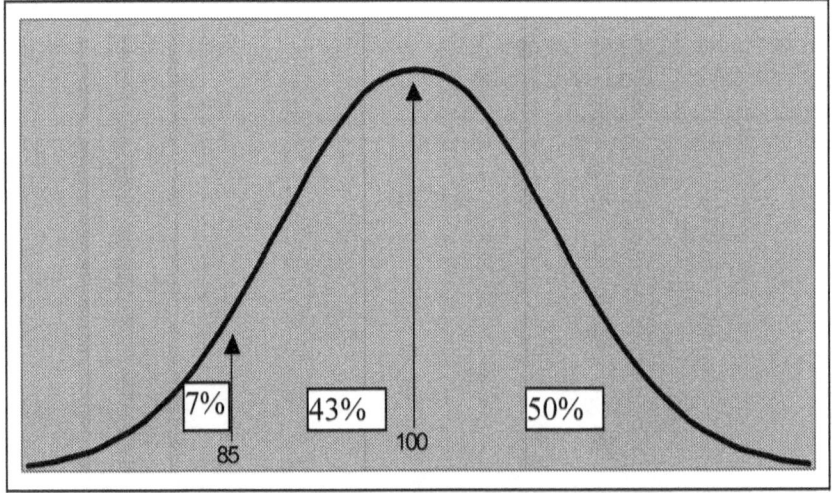

Figure 11

Again, this reasoning works for any normal distribution no matter what the mean or standard deviation. There are many circumstances in nature and in other kinds of measurements where the distribution can be shown to be normal or can be "assumed" to be normal. Given the ease with which calculations can be made, it is no wonder that whenever possible, people will lean toward making the assumption that the distribution is either normal or close enough.

One final word about normal distributions. I have always found it very helpful when solving these types of problems to draw the picture of the normal curve with the area in question shaded in. It helps to think out what values to look up in the table and what additions or subtractions to make.

Binomial Distributions

Another special distribution that is covered in most basic statistics books is the binomial distribution. It applies directly to many of the opinion polls and political surveys that are so popular today, such as how many people favor the latest government policy, or a certain candidate.

The binomial distribution is defined by three criteria:

- A choice between two conditions, such as agree/disagree or heads/tails,

- A fixed number of trials to obtain those outcomes, and

- The probability is the same for each trial. Note that the probability is referred to by the Greek letter pi, π, but don't confuse it with the π used in geometry. It is a probability (with a value between 0 and 1) and not equal to 3.14...

Tossing a coin is a good example of this. The outcomes are heads or tails (two conditions); the coin is tossed a certain number of times to see what happens (trials); and the probability is the same, approximately 50-50, assuming the coin is not bent or dishonest in some other way.

Suppose ten coins are tossed and the number of heads is counted each time. Do this several hundred times and keep track of the number of times 0 heads results, the number of times 1 head results, and so on, all the way up to 10 heads. When the counts that result from these trials are converted to a probability distribution, it is a binomial distribution.

From experience most people know that tossing ten coins will usually result in a mix of heads and tails. Usually it will be close to even, 5 and 5 or 6 and 4, or sometimes 7 and 3. Rarely there will be 8 or more heads or tails. Very rarely there will be 10 heads and no tails, only about 1 in a thousand tries. Although it doesn't happen often, it is possible.

Other applications for the binomial distribution would be sports pools or parts inspection. In betting on sports the outcomes are win-loss, with an assumed probability of 50-50 and a set number of games. The binomial distribution can be used to figure out how much better than pure chance you are doing. Likewise when parts are delivered and need to be inspected for defects, the parts will be good or defective and will likely have a set probability of maybe 98% or 99% based on production systems of the supplier. Even the trucks that deliver the parts will have on-time delivery records of a set probability to let the inventory manager calculate how much of the safety stock is needed to cover for possible late deliveries.

Three points about a binomial distribution

- The mean, also called the expected value, is the probability multiplied by the number of trials (nπ), so for ten coin tosses the mean is 10 x 50% = 5 heads. When tossing a coin 10 times one should expect about 5 heads most of the time.

- The standard deviation can be calculated by taking the square root of nπ (1-π).

- When the binomial distribution is used the statistic is called the **proportion**. It is used to describe the percent of people favoring "x" or the percent of outcomes which are defective, etc. It can equal any probability between 0 and 1, but it must be the same for each of the trials.

Chapter 7 – Discrete Random Variables

Chapter 2 addressed the difference between continuous and discrete variables. Continuous variables have measures that advance smoothly along a spectrum of values. Even though we may describe things like height or time by using individual numbers like 75 inches or 36 minutes, these measurements, time and distance, move smoothly as they change. If you say 36 minutes that means it is closer to exactly 36 minutes than to 35 minutes or 37 minutes. But time flows smoothly between 35 and 36 minutes and can be broken down to seconds, hundredths of seconds, and so on. Continuous variables have measures not only at regular intervals but also at the intervals between depending on how precisely you measure. Distributions like the normal distribution, which are smooth, are usually used to represent continuous variables.

Discrete variables are those that generally are counted rather than measured. How many cars drove past the billboard in an hour. It can be 0, 1, 2, etc., but it will not be 2.17 or 5.9. When the government reports an average of 2.03 children per household, it seems like a funny number. Where does the .03 of a child live? The number of children is discrete, even though the average contains fractional results.

It is not usually accurate to use continuous distributions to describe discrete variables, so slightly different methods are employed. In these cases, however, the concepts of mean, standard deviation, the empirical rule and Chebychev's rule still apply.

Note that sometimes the smallest measurable amount is trivial compared to what is being measured. In those cases the distinction between continuous and discrete fades. When measuring income levels, for example, the earnings can be broken down to pennies only, but when dealing with numbers like $60,000 or $100,000, the pennies would seem to blend together and the data can be treated as continuous.

James Jeray

Chapter 8 – From Descriptive to Inferential Statistics.

Inference is the practice of drawing conclusions about a population from a sample by answering a question about some measurable characteristic, like height, income or opinion. You "infer" that what is true of the sample is true of the whole group.

In the definition of sampling some examples were given of why a sample is often preferred to a census. In some cases it is impossible or impractical to conduct a census – burning all of the light bulbs to find the average life. In others it is just too expensive. In reality a census is rare and a sample is common. But a sample is by definition incomplete; it contains only part of the data required to answer the question. It can be very representative of the population or very misleading.

Inferential statistics provides tools to take the results of a sample and apply those results to the entire population. It relies on some of the statistics discussed previously, especially the concepts of distribution, mean (or proportion) and standard deviation. (Remember, a "statistic" is a measurable characteristic of a sample.)

The key is to begin with the best possible sample, one that is very representative of the population. The problem is somewhat circular. You take a sample because you don't have the time or money to measure the whole population. But because you did not measure the whole population, you can never know how typical the sample is! Variation is once again the culprit. It is not limited. Just as it is possible to toss a coin and get many heads in a row, normal distributions are assumed, in theory, to be infinite in both directions. It is possible, though less and less likely, to find values very far from the mean that are not representative at all of the population. Samples may include these rare and misleading values. The results of sampling are inexact.

This is the main problem of inferential statistics. That is why the results of the sampling are always tempered with considerations like the confidence interval or level of statistical significance. What these terms really are is an admission that the sampler will never really know what the truth is, but thinks he is close, based on the care with which the sample was taken.

The important factors in sampling are the size of the sample and how representative it is. The bigger the sample, the better it tends to be, other things being equal. A common way to strive to make a sample representative is to take a "random sample." This means that the sampler attempts to block out any bias by gathering the sample in a way that the data chosen come by chance and not by some design. College professors were once criticized for sampling among their students, because it was convenient, and then applying the findings to the entire adult population. It can be argued that for many measurements college students are not representative of the entire adult population. They are younger, less experienced and do not have the same motivations as people who never went to college or who went to college at a later time in life. So a random sample is preferred to this convenience sampling.

In some cases samplers try to mimic the population by choosing samples in the same proportions. If the population is 55% female, the sampler may try to "stratify" the sample in the same way. This does add an element of design to the sampling, but it is done in a way that is believed to make the sample more representative. Even though the sample may be 55% women, the next step of choosing the women would be done randomly to once again guard against bias.

Another way to reduce bias is by designing controlled experiments with test groups, control groups and "double-blind" safeguards.

It is extremely important to get a representative sample in order to be able to trust the results. Sampling itself, along with experimental design, are subjects that can be covered in entire books and will not be expanded upon here. The issue of sample size will be addressed briefly in Chapter 9.

Chapter 9 – The Central Limit Theorem

You can't draw conclusions about a population from a sample unless you know how close the sample is to the population. The Central Limit Theorem describes the relationship between the mean or proportion of any sample and the mean or proportion of the population it was drawn from.

Suppose the population has only six observations. Arranged from smallest to largest values, they are called 1, 2, 3, 4, 5, and 6. Table 4 shows the 16 different ways to take a sample of three from a population that has only six members.

123	124	125	126	134	135	136	234
235	236	245	246	345	346	356	456

Table 4

Picture the mean of any of these samples. Since any sample containing 1, the smallest observation in the population, also contains other larger observations, the mean of that sample will be greater than 1. Likewise the mean of any sample will be smaller than 6, because all samples contain some observations smaller than 6. So if you take all means of all possible samples, and arrange them in a distribution, the distribution of these sample means will be less spread out than the original population.

Now look at the number of samples that contain observations from only one end of the population. Only the first two contain mostly small observations and only the last two contain mostly large observations. So not only are the sample means less spread out, they are more clustered around the middle (or the mean) of the original population.

The Central Limit Theorem states this more precisely. It says that given the set of all possible samples "of sufficient size" taken from a population, the mean of the distribution of sample means will be the same as (equal to) the mean of the original population, and that those sample means are normally distributed around the mean of the population. The standard deviation of the distribution of sample means is smaller than the standard deviation of the population by a factor of the square root of the sample size.

This is the key to inferential statistics, because it addresses the issue of how good a sample is. If you take any sample of a given size "n" from a population,

- the mean of that sample will be part of the distribution of sample means.

- that distribution has a mean equal to (or located at) the mean of the population.

- the distribution is spread out less than the population is spread out, and

- it is a normal distribution.

So the mean of the sample is close to what you are wondering about, the mean of the population. How far away it is from the population mean is less spread out than all of the individual observations – you are better off than if you just guessed. How far away you are is generally reduced by taking a bigger sample. Since the sample means are distributed normally, knowing the mean and the standard deviation tells you all that you need to know to go to the table and make fairly precise calculations.

Now this "mean of means" can get really confusing, and it is where some people get lost. All samples have a mean, which is a measurement. If you throw out the rest of the sample and keep only the mean, and do this for all possible samples, you have a whole bunch of measurements (sample means) that can be arranged in a distribution. That group of measurements has a mean, which is equal to the mean of the population. That group of measurements is less spread out than the original population. What the Central Limit Theorem tells you is that the one sample that you picked falls somewhere in this range. You don't know where, but you do know about the range of possibilities.

Consider all samples of size 25 taken from a population of 99. There are more than a trillion-trillion different possibilities. But if you could take the mean of each of those samples and make a new distribution, the mean of the new distribution (of sample means) would be exactly the mean of the original population. It would be less spread out by a factor of 5 (square root of 25), and it would be a normal distribution.

When sampling, although you only get one of all these possible samples, you know about the family that it belongs to. When you take the mean of that sample you can be somewhat confident that it is close to the mean of the population. You also know that it is part of a normal distribution around that mean so you can use the table to calculate some range of possibilities regarding how close it is to the population mean.

The only remaining question concerns the condition that the samples be of "sufficient size." Now this is rather vague. People have done tests to see how small a sample can be and still result in a normal distribution of sample means "no matter what the shape of the distribution of the original population is." These tests have shown that generally samples in the range of 30 to 40 observations are considered large enough. Unless you know that the population is itself pretty close to a bell shape, a sample of around 30 is minimum and bigger is usually better.

This does not mean that 32 or 35 is fine. To get the precision necessary, samples of national public opinion routinely sample hundreds, and often over 1000 people. The 30 to 40 number is used only as the cut-off to determine when special techniques for small sample size should be applied. It is possible to do statistics on these small samples, but the results must be considered less precise and more open to question. Nonetheless, sometimes a sample of only 12 or 16 is all that is available.

Remember, the idea of sampling is to finds out something about a population without having to measure the whole population. The primary concern is to choose a sample that is representative of the population in terms of the particular characteristic being measured. Larger samples will tend to be more representative because they are less influenced by a single unusual observation.

The information given by the Central Limit Theorem has some very practical uses in the real world for using samples to draw conclusions about the whole population. Two specific uses, estimation and hypothesis testing, are covered in the next two chapters.

James Jeray

Chapter 10 – Estimation

How do you find the real mean or proportion of a population when there is not enough time or money to do a complete census? This is a very common issue around election time when the polling organizations try to predict the outcome days, and even weeks before the votes are counted. They do a sample of likely voters and publish the findings about what percent favors which candidate. This is an example of estimation.

In the earlier example of testing light bulbs, a manufacturer wants to make claims about the average life of their bulbs. They can't test all of the bulbs and be left with none to sell, so they are forced to sample and draw conclusions about the average of the population, all bulbs manufactured, based on the average of the sample.

Since any sample obtained will be a member of the family of all possible samples of that size, the Central Limit Theorem tells us that the distribution of this family of sample means has a mean equal to the population mean and a standard deviation smaller than the population standard deviation by a factor of the square root of the sample size. It is also normally distributed. Given this information it is possible to estimate with some level of confidence the true population mean.

Generally speaking, a bigger sample will yield a more accurate estimate of the mean (until the sample grows into a census, and you know for sure). But most of us don't have the time, money or energy to do a census. We must choose a sample in a way that it is unbiased, that is, representative of the population.

The light bulb manufacturer chooses a sample of 100 bulbs. He could choose 200, but then the extra 100 can't be sold. He takes one bulb at random out of each of 100 lots as they are ready to be packaged. The people in the test lab plug them in and keep them lit until they die. Someone keeps track of the hours and calculates the mean hours of life for the sample of 100 light bulbs to be 2000 hours. How close is this to the real mean lifetime for all light bulbs produced?

We know that the mean of all possible samples is the real mean lifetime of the bulbs. We don't really know how good this particular sample is, but it should be close. It is somewhere within a normal distribution of samples where many are very close to the center and a few are further away. If we knew the standard deviation of the population, we could figure out how close by using a normal distribution table found in most text books. But we don't usually know the standard deviation of the population, so we cheat and use the standard deviation of the sample which can be easily calculated from the data. This should give a pretty good estimate of the standard deviation of the population. The standard deviation of the distribution of sample means is 10 times smaller, since the square root of 100 is 10. Using all of this information along with the table, we can develop a range that is likely to include the real average.

Here is what we know. The 2000 hours is close. It could be high or it could be low. Because the means are normally distributed, there is a 95% chance that the 2000 hours is within about 2 standard deviations of the real mean (actually 1.96 from the table). To be on the safe side, let's assume that the sample result is better than the actual and subtract two standard deviations. Remember that those standard deviations are only 1/10 of the population standard deviation. So if the population standard deviation is 200 hours, the standard deviation of the distribution of sample means is only 20 hours. When you buy the bulbs the package may read "average life is 1960 hours." That's 2000 – (2 x 20). It could be that they are understating the average and the bulbs are better than that, but they have reduced the chance of overstating to 5% or less by claiming 2 standard deviations below their sample mean.

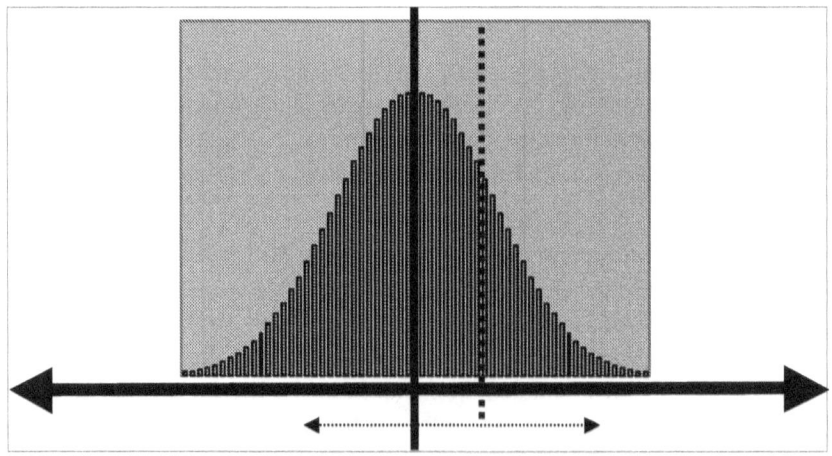

Figure 12

In Figure 12 the range of <u>all light bulb lifetimes</u> is shown by the bold arrow. All possible samples of 100 have means that fall in the normal distribution shown with the mean the same as the population mean. The dotted line is the mean of one sample. If this happens to be the sample chosen, then the sample mean overstates the population mean. But if we calculate 95% brackets around the sample mean, which can be done by using the normal distribution tables, then we can get a reasonable estimate of the actual population mean. In this case, the dotted arrow represents the bracket of (+ or -) 2 standard deviations and shows that it does include the actual population mean.

That is why estimations when correctly reported include the confidence level and the bracket size, "plus or minus." When opinion polls are taken, they are stated as proportions (percentage) plus or minus 3% or 4%. In the fine print it may say something about a 95% confidence interval.

If you wanted to be more sure of your answer there are two courses of action. First take a larger sample, but since the precision is related to the square root of the sample size, you need four times the sample size to double the precision. The other way to be more sure is to increase from a 95% confidence interval to a 99% confidence interval. This will give you a wider range of values (the plus or minus will be larger), but this larger interval will be more likely to include the actual population mean (or proportion).

James Jeray

Chapter 11 – Hypothesis Testing

A hypothesis is a guess about a truth in the world. It is a way of saying, "I think this is true, let's see." The scientific method is built on this process of making a statement and then seeing how the evidence tends to confirm or disprove the statement. The ones that are not disproved are held as true until future evidence comes to contradict them. Then a new explanation is needed.

The variability of the world as was explained in Chapter 1 requires that science use statistics in testing these hypotheses. Whenever you sample, you have an incomplete understanding of the true measures of a population. As seen in the last chapter the sample mean may be very close to the real mean or it may be off by a bit. The bigger the sample, the more precise the answer, but we never know for sure.

Hypothesis testing deals with the results of experiments. In a classic experiment a "treatment" is applied to determine how effective it is. The word treatment implies a medical application, however, in a broad sense it may apply to training people, applying a fertilizer to a garden, or a host of other situations. (There was a news story in March 2002 that reported on a study saying that chewing gum increases people's cognitive skills. In this case the gum chewing was considered the treatment.) The main point is that a treatment is applied to get a result, a result that is different from the no-treatment result. The treatment costs money (or time or effort). So you don't want to spend the money on it unless it really works.

The primary question becomes, whether the treatment works better than doing nothing. This would be an easy question to answer if it were not for the variability in the world. Everything varies, so doing nothing will produce variable results and applying the treatment will produce variable results. For example, give a new cancer drug to a group of patients, some may recover quickly, some slowly, and some not at all – variability. Do nothing for a similar group of cancer patients, some may recover quickly, some slowly, and some not at all – again variability. How many more patients need to recover as a result of the drug in order for the drug to be declared effective? That is a question for statistics, specifically in the area of hypothesis testing.

Usually a sample is taken from the treated group and compared either to data on the population as a whole or to measurements on a "control group." Statistics is used to determine if the result is different enough to conclude that the treatment had a positive effect. The results of the treated group and the untreated group both are variable, so they each have a mean and standard deviation. Figure 13 shows a picture of the two results where the mean of the treated group is better, but there is a lot of overlap. Figure 14 shows a much wider difference. It is clear that results like those in Figure 14 would more easily persuade someone of the effectiveness of the treatment, but there must be standards. The Central Limit Theorem states that the means of all samples are normally distributed. This allows the probability of any size separation, comparing the treatment to chance variation, to be calculated.

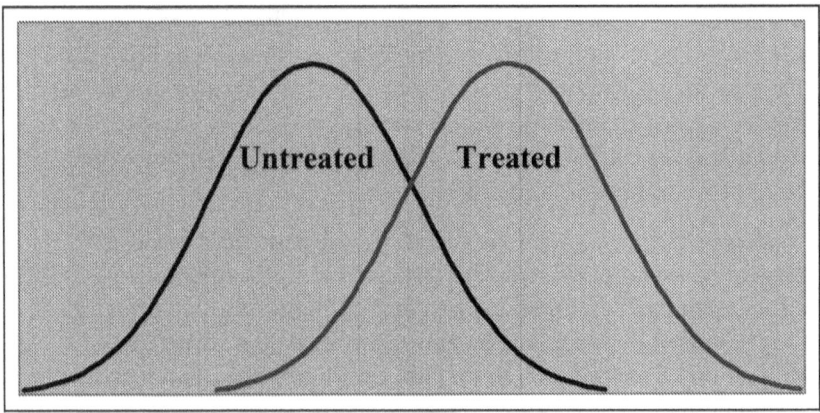

Figure 13

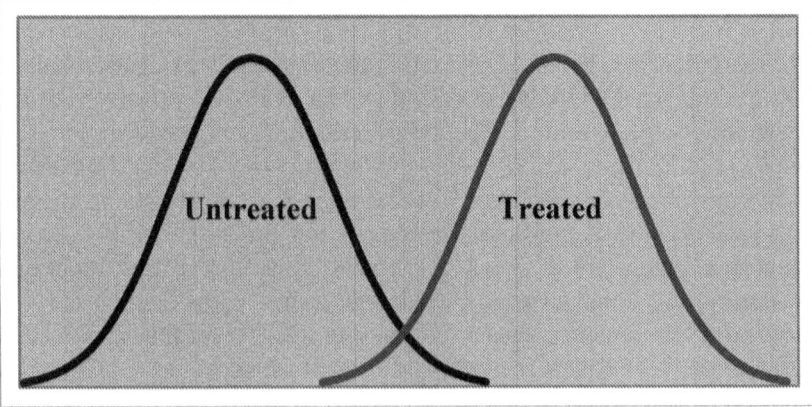

Figure 14

This is where the concept of "statistical significance" comes in. It establishes the standard. Results are said to have a significance level of 95% or 99%. The latter is more persuasive than the former. These results, however, can never absolutely prove that a treatment has the desired effect. Variability and sampling limit the precision of such tests. Even in Figure 14 where the results appear to be clear cut, it would be possible to draw samples such that the treated group looks inferior to the untreated group. Likewise, it would be possible to draw a sample for an ineffective treatment showing statistically significant evidence of effectiveness. Both cases lead to wrong conclusions.

For those who are taking or who have taken basic statistics courses, this is why there is such an emphasis on the "null hypothesis," the hypothesis that there is no difference between the treated group and the untreated group. You may have enough evidence to reject the null hypothesis and conclude that the treatment does produce a difference, but you can not be absolutely sure.

Chapter 12 – Other methods

The following methods are often touched upon in basic statistics courses and are handy for their particular uses. Here is a very brief summary.

Chi-square

Sometimes the data is gathered by categories to determine if there is a difference between those categories with respect to some outcome. Do people who take the same side of the argument on posting of the 10 Commandments in schools or on carrying concealed weapons belong to the same political party or is there no connection? Do men and women have the same favorite spectator sports or are there apparent differences? The categories in the above examples are the party affiliation or sex of the individuals.

Here is an example. There is a belief that men around 40 years old, go through a mid-life crisis and buy a sports car. To find out if men in their 40s are a hotter market for sports cars, an automobile maker might gather data as shown on the left side of table 5 below. In the table on the right are the calculated expected values, that is, the answer in each category that would make the proportions exactly the same between the age groups. The question is whether the numbers in each cell on the left are enough different from the corresponding numbers on the right to make one suspect that the results might be caused by something other than random variation.

Age	Sports Car	Truck / SUV	Other	Age	Sports Car	Truck / SUV	Other
20's	6	11	13	20's	4.5	9.2	16.3
30's	5	10	26	30's	6.1	12.6	22.3
40's	8	17	25	40's	7.5	15.4	27.1
Older	2	5	12	Older	2.9	5.8	10.3
Totals	21	43	76	Totals	21	43	76

Table 5   Actual Data   Expected Values

In this example the 8 sports cars bought by the men in their 40s is a little higher than, but pretty close to the 7.5 "expected" number. Is it high enough based on the sample size to draw a conclusion of preference by these men or is it just the variation that would show up in any sample?

The Chi-square technique is used to answer this question. It compares a difference calculation for each cell against a distribution called the X^2 (the Greek letter Chi). It tells, as in hypothesis testing, the percent to which the difference is statistically significant. If the percentage is high, around 95% or higher, it may be assumed that there is more at work than mere random variation. Otherwise, no conclusion can be made.

ANOVA

Analysis of variance (ANOVA) is a technique used to compare the means of multiple sets of variables at the same time. The question remains the same, "Is there a difference between the means that can be attributed to outside factors or is it just the result of random variation?"

Experimenters may be comparing the measured performance of three different brands of automobiles. They may be comparing the effectiveness of four different drugs applied to different groups of patients with the same illness.

Conditions required include that the samples being tested come from normally distributed populations and that the standard deviations be homogeneous, that is, approximately the same. The test statistic is the F-statistic, which gives a probability allowing the experimenter to decide if he will accept the statistical significance of the difference or say that the difference is not strong enough to rule out random variation as the cause.

Correlation and Regression

So far the topics have covered what are known as univariate analysis. This means that only one variable at a time is looked at: height, age, income, etc. When looking at the mean and standard deviation it is calculated for only one variable. Estimation and hypothesis testing were only interested in one variable at a time.

There are two methods that are often included in basic statistics courses that involve two variables. These are correlation and regression.

Correlation talks about the (linear) relationship between two variables. Does one increase as the other increases? Does one increase as the other decreases? To what extent do they vary together? For example, the color of a star is related to the age of the star. The heights of people and their weights are generally correlated positively. That is, taller people tend to be heavier – there is more of them. This is not always the case because everything varies and you find short heavy people as well as tall thin people, but in general it is true. The amount of snowfall may be correlated with the sales of a sporting goods store between the months of December and March.

Correlation is expressed numerically between −1 and +1. A correlation of +1 means that variables are perfectly correlated, that an increase in one leads to an increase in the other that is absolutely predictable, always the same ratio. A correlation of 0 means that there is no linear relation between one and the other. (There may be some relationship; it just can't be expressed as a linear function.) If one increases the other will sometimes increase a lot, sometimes increase a little, and sometimes decrease. A correlation of −1 means that as one increases by a certain amount, the other decreases by a certain (not necessarily the same) amount, but like a correlation of +1, the decrease is always the same for a given increase. A correlation of .90 is usually considered very high. The two variables move together, up or down, not exactly in sync, but very close. The opposite is true of a correlation of .10. It is closer to no correlation at all. The rest of the values, .25, .48, .79 or any others are in the middle in terms of how closely the two track together.

So the sign, + or -, shows whether they move in the same or opposite directions, and the value shows how consistently the change of one is relative to a change in the other.

An important thing to remember is that correlation is a mathematical concept. It has nothing to do with cause and effect. To say that variables are correlated to a high degree only says that the sample of values is mathematically consistent in their common movement. It says nothing about whether one causes the other or whether they result from a common cause. This point is stressed in every statistics book, but people still try to use correlation to try to prove causality. That as air pollution goes up the cases of asthma go up may be true, but it is not by itself proof that one causes the other. More information is needed.

The other common bivariate method is called linear regression. Using this method the values of two variables are plotted on a graph and the best fit line is drawn. This line is assumed to show the slope of the relative change in correlated variables. If the slope of the line is 1, then as one increases by a single unit, the second tends to increase by a single unit. So, if you had a number of examples of related pairs of variables that established such a pattern, you could estimate the value of an unknown variable given the value of the other half of the pair.

Suppose you had a list of results for an elementary school fitness test for boys running 50 yards. In general the older boys would be running faster.

There would be a negative correlation between age and the time it took, that is, as the age went up the running time would go down. There would also be variation; all of the points would not line up perfectly. A PC could be used to calculate the best fit line. (The distance between any point and the line is called the error and the best fit line is the one that minimizes the total error.) Given that information and the age of an untested student, it would be possible to predict within a range how fast that boy would run 50 yards.

The dangers with regression arise from extrapolation and from the assumption of linearity. Regression should only be used to make predictions within the range of known data, interpolation. There is danger in using it to guess at points beyond the range of known data. In the above example, you wouldn't want to use the same regression line to predict the running time of a boy in high school. The age is out of the range, and this extrapolation increases the chance of an error.

This is an error commonly made when trying to predict the future. Data exists for the past year, but there is no data yet for next year. To predict that the trend line will continue is a pure case of extrapolation. That is why mutual funds advertise that "past performance is not necessarily a predictor of future returns." This also explains why forecasting (economic, business, or weather) is so difficult.

Second, the "best fit" is not always a straight line. Sometimes it is a curve and the linear method will yield good results only in a very small range where a line is close enough.

Figure 15 shows a set of data that is positively correlated with a correlation coefficient of .90. The "best fit" regression line is drawn in.

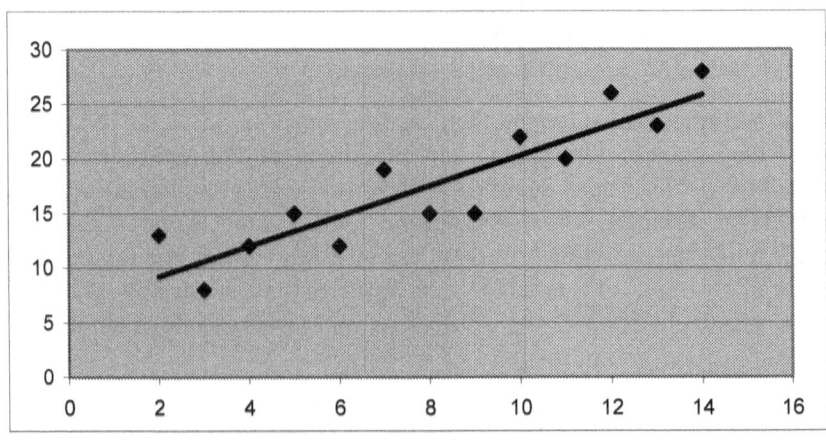

Figure 15

Multivariate Statistics

Advanced courses in statistics deal with analysis methods that can consider several variables at one time. Some of these are multiple regression, canonical correlation, discriminant analysis, or factor analysis. These are much more complex, and applications are less common. One example is in economic forecasting where changes in the future may be related to several variables in the present. Another is the development of psychological tests where the answers to several questions become the variables used to determine the level of a single personality trait.

James Jeray

Summary

Statistics is all around us and touches on our daily lives, so like it or not, the fact that it can be intimidating should not be an excuse.

The mathematics of basic statistics is not very complicated. It is little more than some simple algebra and arithmetic. In any case computers and calculators can do statistical calculations that are far beyond the methods discussed here. A simple spread sheet program can calculate the average and standard deviation of a large set of observations very easily.

So the scary part of statistics seems to be the vocabulary and the underlying theory. It has been the intention of this short book to explain the basic concepts and to introduce the terminology while leaving out the detailed calculations and the exercises. To keep it short, examples were limited with the hope that a few very clear and simple examples were enough to communicate the main ideas behind the subject.

There are several main points in closing.

The variation of the world makes some very unusual things happen. Because of this it is really impossible to use statistics to prove anything without some doubt. It is unusual to roll 10 sevens in a row at a craps table, the odds are 1:165,000,000 against it. But it can happen, and when it does, it does not prove that you have psychic powers or that the dice are loaded. This is important to remember when hearing a news report of a single study that indicates new beneficial effects of certain vitamins or that a new product can reduce the length of a cold or the flu. Despite the fact that they are honestly and professionally done, they can still lead to the wrong conclusion, because of the random variation everywhere. So one study contradicting another is not only possible, but to be expected from time to time. Before accepting these results, look for multiple studies done by reputable groups, which have been reviewed and endorsed by other professionals. Before drugs are approved, the FDA has standards that must to be met, likewise the EEOC when discrimination is alleged, and still mistakes are occasionally made.

When faced with any of this information whether it be on the news or as part of your job, there are certain details that must be satisfactorily addressed. These include how big was the sample; was it of sufficient size; were good sampling techniques used to avoid bias; are they trying to apply the result to a population that goes beyond the sample; are conclusions of causality being drawn from data that shows only correlation; and do the people conducting the polls or studies have a stake, financial or otherwise in the outcome?

Finally in statistics, interpretation is far more important than calculation. It doesn't do you any good to get the right answer but not be able to tell what it means (or what it doesn't mean). Any PC can do the numbers. The brain work is to figure out where the numbers take you and what new insight or conclusions about your job or about your world can be fairly drawn. Because of variation, it's never going to be a sure thing.

Appendix – Sample Normal Distribution Table

This table gives the percentage of the area under the curve from the mean in either direction for given z-scores. The way the tables is laid out lets you look up z-scores to two decimal places. The first column on the left shows the z-score to one decimal place and the other nine columns across allow for the next decimal place.

The cell at 1.0 in the first column (bold) shows where the (approximately) 34% in Figure 10 comes from. Likewise, the cell at 1.5 in the first column shows the (approximately) 43% used in Figure 11. A z-score of 1.96 shows .4750, or 47.5%, so the range from +1.96 to −1.96 has a probability of 2 x 47.5% = 95%

z-score	0.00	0.01	0.02	0.03	0.04	0.05	0.06	0.07	0.08	0.09
0.0	0.0000	0.0040	0.0080	0.0120	0.0160	0.0199	0.0239	0.0279	0.0319	0.0359
0.1	0.0398	0.0438	0.0478	0.0517	0.0557	0.0596	0.0636	0.0675	0.0714	0.0753
0.2	0.0793	0.0832	0.0871	0.0910	0.0948	0.0987	0.1026	0.1064	0.1103	0.1141
0.3	0.1179	0.1217	0.1255	0.1293	0.1331	0.1368	0.1406	0.1443	0.1480	0.1517
0.4	0.1554	0.1591	0.1628	0.1664	0.1700	0.1736	0.1772	0.1808	0.1844	0.1879
0.5	0.1915	0.1950	0.1985	0.2019	0.2054	0.2088	0.2123	0.2157	0.2190	0.2224
0.6	0.2257	0.2291	0.2324	0.2357	0.2389	0.2422	0.2454	0.2486	0.2517	0.2549
0.7	0.2580	0.2611	0.2642	0.2673	0.2704	0.2734	0.2764	0.2794	0.2823	0.2852
0.8	0.2881	0.2910	0.2939	0.2967	0.2995	0.3023	0.3051	0.3078	0.3106	0.3133
0.9	0.3159	0.3186	0.3212	0.3238	0.3264	0.3289	0.3315	0.3340	0.3365	0.3389
1.0	**0.3413**	0.3438	0.3461	0.3485	0.3508	0.3531	0.3554	0.3577	0.3599	0.3621
1.1	0.3643	0.3665	0.3686	0.3708	0.3729	0.3749	0.3770	0.3790	0.3810	0.3830
1.2	0.3849	0.3869	0.3888	0.3907	0.3925	0.3944	0.3962	0.3980	0.3997	0.4015
1.3	0.4032	0.4049	0.4066	0.4082	0.4099	0.4115	0.4131	0.4147	0.4162	0.4177
1.4	0.4192	0.4207	0.4222	0.4236	0.4251	0.4265	0.4279	0.4292	0.4306	0.4319
1.5	0.4332	0.4345	0.4357	0.4370	0.4382	0.4394	0.4406	0.4418	0.4429	0.4441
1.6	0.4452	0.4463	0.4474	0.4484	0.4495	0.4505	0.4515	0.4525	0.4535	0.4545
1.7	0.4554	0.4564	0.4573	0.4582	0.4591	0.4599	0.4608	0.4616	0.4625	0.4633
1.8	0.4641	0.4649	0.4656	0.4664	0.4671	0.4678	0.4686	0.4693	0.4699	0.4706
1.9	0.4713	0.4719	0.4726	0.4732	0.4738	0.4744	0.4750	0.4756	0.4761	0.4767
2.0	0.4772	0.4778	0.4783	0.4788	0.4793	0.4798	0.4803	0.4808	0.4812	0.4817
2.5	0.4938	0.4940	0.4941	0.4943	0.4945	0.4946	0.4948	0.4949	0.4951	0.4952
3.0	0.4987	0.4987	0.4987	0.4988	0.4988	0.4989	0.4989	0.4989	0.4990	0.4990

James Jeray

About the Author

James Jeray directs an organization which uses mathematics to model business processes. Ensuring that practical solutions are delivered to customers means that technical issues must be translated into everyday language. Communications is key.

These talents were crucial when he was asked to teach evening classes in statistics to undergraduate business majors. These students were not thrilled at the prospect of this required course. Soon there were compliments from both students and administrators on his ability to find everyday examples to clarify and de-mystify the concepts before teaching the mechanics of calculation. This experience and the encouragement of colleagues inspired the writing of <u>Statistics: Short and Simple</u>.

Mr. Jeray holds a BS in Engineering, an MS in Administrative Science and an MBA.

www.ingramcontent.com/pod-product-compliance
Lightning Source LLC
Chambersburg PA
CBHW021909170526
45157CB00005B/2026